Concerned Citizen
Altruistic Endeavours

Protector

Lincoln Coull

Copyright © 2024 (Lincoln Coull)
All rights reserved worldwide.

No part of the book may be copied or changed in any format, sold, or used in a way other than what is outlined in this book, under any circumstances, without the prior written permission of the publisher.

Publisher: Inspiring Publishers,
P.O. Box 159, Calwell, ACT Australia 2905
Email: publishaspg@gmail.com
http://www.inspiringpublishers.com

 A catalogue record for this book is available from the National Library of Australia

National Library of Australia The Prepublication Data Service

Author: Lincoln Coull
Title: Concerned Citizen - Altruistic Endeavours
Genre: Non-fiction

Paperback ISBN: 978-1-923087-36-1
ePub2 ISBN: 978-1-923087-35-4

**TO THE PROTECTION AND PRESERVATION
OF MOTHER EARTH – and all therein**

Please allow me to share some of my experiences that you'll hopefully find interesting and possibly inspiring.

Contents

Growing up 1
Save Our Culture..5
Gore..9
Save Mother Earth 10
The Commission 21
Nonchalant Heroes..23
A New York Treat 25
The Flag 27
Madam 29
Idealism..32
Spiritually...35
How we got here 37
The Contemporary Anthropocene era 44
Positive Reflections and Solutions...........................47
We may have turned the corner................................52
From my stance 57
A final thought..63
Notes 67
Index 70

Growing up

Growing up in Adelaide in the 70's, life seemed freer, simpler and comfortably raw. There was no Internet, no manic political correctness and no Big Brother. I was raised in a typical working class, tough, left neighborhood where most everyone was true, straight forward and genuine. You had to stand your ground and keep the ruff necks on side. When my twin brother and I were 5 years old our parents divorced. My Dad, who worked carting timber on his truck off the wharfs to the company storage depot and various places around Adelaide, took my brother and I, every other weekend, fishing and camping. In later years I was so thankful he did, as later on, it led me to travel further. We still go fishing together once or twice a year, either in his boat or on a beach or jetty somewhere. But it was Mum who raised us both, mostly by herself, unselfishly sacrificing her own life to feed us and keep a roof over our heads. Money was always tight then and there were times where we went without.

Most people took their hat off to my Mum for how much of a good job she did in raising us.

When I was about 12, Mum asked my brother and I to come in from playing in the back yard. She decided it was time to tell us about the constant sexual abuse she and her sister endured growing up as children. Admittedly, this had a certain effect on my psyche. For years it pained me to know how much she had suffered. But more so, it was hard for me to live with the fact that I couldn't do anything about it. I was powerless to help my Mum. As intelligent and well-read as she is, I saw how that experience hindered and limited her life. For many years, she worked, shopped, went to church, read and stayed in the safety of her own home. She never ventured out, never travelled or pursued other potential life enriching endeavours.

But somehow, to her credit, that seemed to be compensated by the fact that she had, naturally developed and ever readily utilized a certain innate counselling ability. She willingly and selflessly gave herself up to listening and counselling many, many people of all ages over the decades. All of my mates and family members found it naturally easy to share almost everything with her. They visibly felt better in themselves after having spoken with Mum. Although she did a semester or two studying Psychology, it was more an innate understanding and compassion to help others that she had and generously gave. Thankfully, she is leading a happier, more fulfilling life now in her retirement years. Ever thankful God bestowed her as my Mum. I think some of that innate caring and understanding filtered down to me.

Then when I was about 14, I remember these two guys from Mum's church visited our place on a Saturday afternoon. One of them, who looked a bit hippy, claimed he could see people's Auras. Sitting around the dining table, somewhere in the conversation, he looked at me and asked, and what do you want to do? My reply, innocently, was that I wanted to do as much good for the world as I could. He responded saying wow, look at his Aura light up!

Like most kids being full of beans, I didn't like school much as having to be in a classroom for most of the day felt restrictive.

By the time I got to my third year in high school, going to school was more of a social event. I didn't spend much time on doing schoolwork or study, it was more a case of hanging out with my mates, doing sports, chatting up girls and working out what we were going to do on the weekend. Luckily, as I had the attention of the Career advisor at school, he helped me get an interview and the job straight after finishing my fifth year. I thought I would never go back to school again.

I proceeded to work for the Vehicle Builders Union as a clerk in the city. At first, I thought it was great, as most of my best mates also had jobs in the city, we would have lunches together and partied every Friday and or Saturday night on the town. Relatively though, that job exposed me to workers' rights and how to stand up for what's right. Moreover, as there were a constant stream of Lawyers and Politicians floating through there, I became aware of how they and their decisions significantly induced change. Hence, it was also where I started to become politically aware.

One morning, one of the lawyers who knew mine was a dead-end job, made the effort to advise me "if you really want to go anywhere or advance in a career, you will have to do some type of further education and study." I think it was about twelve months later after having been in that position for 3 years, I left to go back to do what I never thought I would do, my adult Matriculation year and then apply for University.

I passed Matriculation and was accepted into Flinders University. In 1990, I showed up to Uni having no idea of what courses to study and found myself enrolling in Politics as my Major, with English Literature and Asian studies minors. I should also mention that part of my Political studies involved two semesters of Environmentalism.

Before University though, I thought I knew it all, but then soon after starting my studies, I came to realize just how little I did know; how ignorant I was to the world. Uni opened my eyes and while I experienced some of the most amazing times socially there, I progressed to becoming more Politically and also Environmentally aware on a national and global scale. I also remember learning how Democracy had initially begun. In ancient Greece, all the people of the village, regardless of their status, would gather together in an open forum to voice their opinions on what they thought should be done regarding any particular matter. In those times, usually it was the person who had the loudest voice, who influenced the many to make a decision.

Save Our Culture

After University, I had a myriad of different jobs from bar tending in restaurants, to working in factories and on building sites, to ending up back in an office job in the city – and all the while I kept an eye on the Australian political situation. Through the 90's, I remember realizing how the Australian Government and Corporate sectors seemed to be blindly following the American capitalist work ethic of making the front liners work at breakneck speed and performing multiple different tasks simultaneously. I was convinced that our Australian society was gradually descending to their "dog eat dog" lifestyle; which was eroding our essence, our "Mateship" ethos.

I remember one day, talking to a woman at the checkout in the local supermarket, who told me she went to work over in the US and then came back to Australia for the very same reason. This inspired me to act, to do something about preserving our culture. I wondered how or what would be the best, most effective way of doing this. As I wasn't a billionaire, rock star or politician, I didn't have the resources they had to make a significant impact. Relatively, while thousands of people send one off letters to the ombudsman or to the editor of a newspaper, none of them ever sent the same letter out to the masses. At this time, the computer and internet had firmly started their systemic takeover of the world and it became evident to me that the only real way of being heard, to really have an impact and effect, was to contact as many people as I could via email; to have the loudest voice.

I proceeded to embark on an email address gaining mission. Each weekend for several months I spent countless hours accumulating email addresses from every place I could think of in Australia. I copied and pasted them from every newspaper, every journalist and editor, every radio and tv station and current affairs type programs. I also attained the email addresses of every prominent Australian personality and actor I could think of.

Also at that time, I noticed the then Liberal Government was proposing to implement the new "IR laws", which would give employers great power and could significantly reduce the pay and rights of workers. I was convinced it would further erode our society to the point where, like in America, a lot of Australians would end up having to work two or three jobs just to get by.

I decided then to also grab almost every minister's email address in the Labor Party, the Greens and the Independents as well. When I was satisfied that I'd attained enough, had the desired reach, I forwarded the following article on the 20th of May, 2007:

20/05/2007 – emailed the following article to most every Australian Newspaper, TV and Radio Stations and the majority of Ministers/Senators in the Labor Party, The Greens, The Democrats and some Independents.

Save our Culture

May I express my thoughts on the new Industrial Relations legislation the Liberal Government is proposing.

We all know of our famous Australian attribute of knocking the person who does better than others, but we also possess that valuable, cultivated convention of supporting and looking after each other, which has been entrenched by our "Mateship" ethos. However, in the last five to ten years the Australian work ethic has changed and accordingly, the workplace attitudes of Australians has also significantly changed; to the point where I have seen a general acceptance of the denigration of one another grow and this is taking the "one up" vernacular to un attractive levels. Rest assured, this new legislation will only increase this phenomenon more so.

If this legislation is passed as law, the Australian attitude will worsen here, as we will be forced to compete against our workmate/colleague and neighbour for our very livelihood. Sadly this will filter through to our social lives and our view toward each other will gradually be moulded toward fighting each other both psychologically and physically. Indeed like in America, it will encourage a "dog eat dog", heartless, every-man-for-himself attitude that will not benefit but poison our great lifestyle. Violence, theft and drug trafficking will surely rise to unprecedented heights. I am certain this new legislation will erode the very core of what it is to be Australian.
Infact, it is unAustralian.

Thus, is this "dog eat dog" attitude really what we want?

Concerned citizen,

Lincoln Coull
Seacliff SA

Not long after, famous Australian celebrity, the late Bert Newton came out with a new "20 to 1" TV program that show cased and emphasized all that was great about Australia. I remember the huge nationwide rallies that followed and unfolded in every Capital city protesting against the IR laws and a couple of commercials started airing on TV, encouraging everyone to look after your work mates. To my relief, I felt elated to later see the Labor Party win the next election in November that same year and subsequently abolish the IR laws.

A few months later into the next year, I remember viewing on line one of the famous Sir David Attenborough documentaries that he had made in 1984, where he mentioned for the first time, the Carbon Emissions threat.[1] Then while walking along the beach reflecting on the then current Environmental situation around the world, I realized that, besides Mr. Gores' gallant efforts, nothing really was being done to combat Climate Change and I felt compelled to do something.

Even though I had studied Environmentalism at University, I still wanted to access all of the latest information that I could on Climate Change. So, for the next 6 months I resigned all my weekends to sitting in front of the computer earnestly joining every Green movement and body I could find, from the obvious Green Peace to the Sunrise Movement, Climate Ad Project, Future Earth, Environ News, Climate Central, ICCCAD, CICERO, IPCC and many more. More importantly though, I began to again look up and copy every email address I could find across the globe. I grew vast lists of emails from most every newspaper, magazine and politician in the UK, Japan, Canada, France, China, USA, Russia, Holland, India and Germany. In the end I had amassed over 4000 email addresses.

Gore

With the utmost respect and support for the former Vice President Mr. Al Gore and his indisputable, great efforts - mindful that since his early years in the late seventies, he was tenaciously putting out the message about Climate Change. You'd remember his famous documentary "An Inconvenient Truth" in 2006 and he published subsequent books on the subject.[1] However, he really wasn't achieving the traction he was striving for. The reason for this was because, for some reason, he did not inform the masses of the most obvious fact; but I subsequently did. I determinedly began my own mission to show the world that Climate Change wasn't something that was to happen in the future, it was evidently happening right now.

Save Mother Earth

I patiently waited and looked every week at the weather going across Mount Kosciusko, the tallest mountain in my country. I calculated how long it would take me to drive there, how much it would cost in fuel and what provisions I needed to take to accomplish my mission. Having driven through that way a few years earlier when I was coming home from a six-month secondment in our nation's capital Canberra, my familiarity gave me the confidence to tackle this journey once more. When the window of three good days of weather opened, I seized the opportunity. As funds were a bit tight, I made up sandwiches and drink, raided my coin collection and set off early in the morning on the following 2,500 kilometre round trip journey in my four-wheel drive wagon. Along the way, I stopped twice for fuel and food and arrived at Lake Jindabyne by mid afternoon. It was about 45 minutes' drive from the mountain. I drove around and found a nice place overlooking the lake to camp for the night. I welcomed the couple of hours to relax after the long drive and contemplated my task set for the next morning. I ate and drank and set my phone alarm for half an hour before sunrise. Having taken out the back seat in my wagon and replaced it with a four-inch foam mattress length ways, it was quite a cozy sleep.

In the morning I rose, had a quick bite and drove to the Summit walk near Charlotte Pass. Arriving at the car park which was at the start of the Summit walk, I was thankful the weather was still good. I put my thermals on under my jeans, my prized (expensive) Columbia Vertex Mountain jacket on over my jumper and my loaded backpack on my back. I was ready and having started my trek, I was quietly pleased to also have my favourite, worn in, blue Doc Martin six-hole boots on. I had previously worn them when I climbed Uluru a few years earlier. However, after about an hour into my 18 kilometre walk, I began to feel a slight pain on the bottom of both my feet. I soon realized I had started to form blisters and was baffled as to how, as this had never happened before and I found myself asking, why now? I became very frustrated facing the decision of whether to turn back and drive all the way home having not accomplished what I'd set out to do or continue on for the next 7 or so hours with my blisters. I'd come so far, I chose to go on.

Later, about two thirds into the walk, I decided to rest and have something to eat and drink at Seamans Hut; a small, very old and basic stone shanty. I met a dad and his son doing the same walk. I got along well with them and it comforted me to know the dad was Scottish, as my heritage is Scottish on both sides of my parents. We proceeded to walk and chat with each other along the rest of the way and reached the Summit together.

At the top of the mountain, after putting up my flag and having taken some photos and a short video, I sat and prayed for us to Save Mother Earth and for God to look after my mother and brother and for no benefit to come to myself. I did this, thinking it was the noblest thing I could do.

A few hours on, I made it back to the car park, opened the rear door of my wagon and peeled off my jacket. Exhausted, I sat down in the back of my car and took off my boots to see blisters the size of my thumb dotted over the soles of both my feet. I then drove for the next 6 hours, found a place to camp overnight next to the Murray River and then arrived back home to Adelaide before noon the next day.

Shortly after, I edited the short video I made, posted it on YouTube and then emailed the below articles out across the globe:

April 2008 - I have sent the following article to most every newspaper in the UK, Japan, Canada, France, China, USA, Russia, Holland, India and Germany.

You know I drove 2500 kilometres and climbed 2200 metres to put this flag up,

SAVE MOTHER EARTH

Who amongst you has the courage, who amongst you is humane enough to meet this challenge?

There's a sacred flag on the tallest peak in the world on Mount Everest, and now there's one on the tallest peak in Australia, Mount Kosciuzko; both saying SAVE MOTHER EARTH.

I challenge you to set your flag on the tallest peak in your land and make this issue of Saving Mother Earth the highest of importance in your country.

SAVE MOTHER EARTH…………SAVE MOTHER EARTH !

Yeah, help Save Mother Earth.

Lincoln Coull
Adelaide, Australia

PS: As confirmed on my 50 second movie on You Tube titled, you guessed it
'SAVE MOTHER EARTH'.   http://www.youtube.com/watch?v=brj0_KuCw68

April 2008 I also sent the following two articles to most every Australian Newspaper, TV and Radio Stations and the majority of Ministers/Senators in the Labor Party, The Greens, The Democrats and some Independents.

SAVE MOTHER EARTH !

Seventeen years ago I studied a couple of semesters of Environmentalism, when the subject was still in it's infancy and largely conjectural. Now like most of the scientists today, I am convinced that climate change is occurring as a result of human interaction.
The evidence on a global scale is too exponential now.
As for what I'm doing to save the planet? Well I can only do as much as my income can allow. I changed all the light bulbs and turn everything electric off when not in use. I also recycle my household water and my trash. Although, if I could afford it, I'd have the whole of my roof covered in solar panels connected to four giant batteries and also attach two windmills at either end; to acquire enough power to get completely off the grid. Then either dig a bore and/or as I live by the sea, buy my own desalination device.

I think we'd all agree to the individual having to do her or his little bit, but while I don't have the power of Mr. Gates, Murdoch, Bono or Gore, I am utilising the resources within my means. As the benevolent Robert F Kennedy Jr has quoted "it was imperative governments stopped bankrolling big industrial polluters by providing millions in subsidies". "If the market place was level and subsidies to oil and coal companies were scrapped then vital renewable energy sources such as solar power, could compete and prosper".

You see, fundamentally we need to encourage a global collective to put pressure on the bigger players that can and need to induce the necessary change.

I couldn't go on without having done something.

Time is short, the Window small....

CONCERNED CITIZEN - ALTRUISTIC ENDEAVOURS

July 09 This email sent to most all Canadian,Japanese,UK,China,USA,French,Russian,Holland,German, Indian and Australia newspapers. And now (21/11/09) most every Senator/Minister and Presidents/Prime Ministers in those countries.

Dear Senator/Minister,

You know before the ancients, it took the indigenous tribes of every continent thousands of years to acquire the knowledge that we have to share the sacred responsibility of the caretaker role; to ensure the perpetuation of our species.

Have you ever considered the irony of how the two biggest global issues currently, are the collapse of the financial system and the climate crisis? Money vs Mother Earth. The market machine falls, is attended to and will always recover; not so with the climate, for without the necessary action, we won't recover.
Perhaps now we might appreciate the seriousness and immediacy of it.

There will be some financial suffering, although minimal by comparison, as it's necessary to prove to the landlord we want to keep on paying the rent.

For a more enlightened and inspirational view on the path we must take when addressing climate change, prey watch the video of the Australian Senator Christine Milne's speech at: http://vimeo.com/5282859

It seems the dawn of a new era is upon us.

You know, we built the Pyramids, the China Wall and put humans on the Moon;
it's time for us to outdo ourselves once again.

Do us and your nation proud, ask your government to vote the necessary 40% reduction in emissions by 2020 and Zero emissions by 2050 at Copenhagen.

It's on you Now,

Time is short....The window ever smaller ....   Stand tall.

Yours truly,

Lincoln Coull
Concerned
Global Citizen

PS: Innately, we all want to live in a clean world don't we?

After having sent out the above emails in April 2008 and July 2009 to various media, journalists and editors, governments, politicians and Leaders in ten of the major countries around the world, I then sent the below, more specific emails in April 2011 to every corner of the US including every government department relating to the environment and energy, of which, there were more than I anticipated. I know I caught President Obama's attention.

I also sent the same email to every Australian email address again. This email however, included graphs and information evincing that the effects of Climate Change, that scientists had been predicting, were actually happening, indeed occurring now:

10/4/2011 - This email sent to most every American Newspaper, Radio and Television Stations. And now most every Minister, Senator and the President.

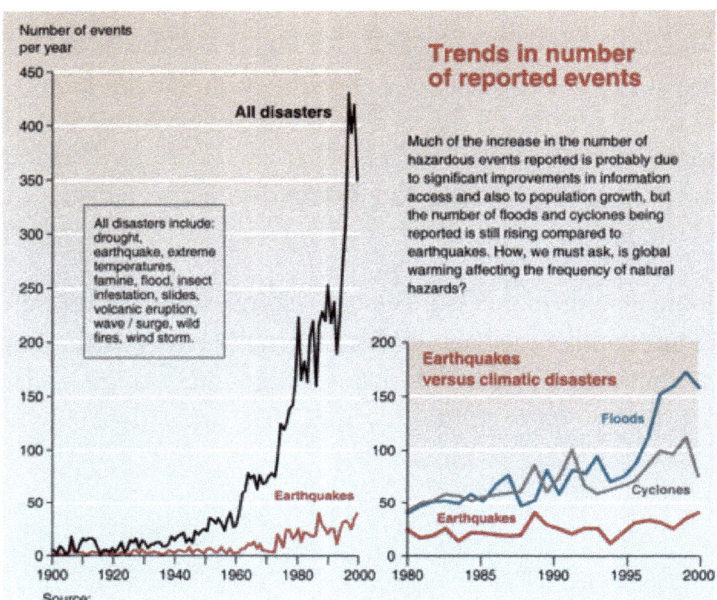

Natural Disasters on Increase – Insurers confirm

Since 1999 the insurance companies have been confirming the increase in frequency and severity of Natural Disasters. "killing and injuring many millions of people every year and causing mounting economic losses". Klaus Topfer UN Environment Programme (UNEP) director, former German environment minister.

"insurance companies are calling it a "**catastrophe trend**." - Discover Magazine, Vol. 21 No. 6 (June 2000)

BERLIN — Insurers' losses from natural disasters rose by about 50 percent in 2008, with **climate change** increasingly a factor, a leading reinsurer said. Munich Re said in an annual review that insured losses came in at $45 billion this year, up from nearly $30 billion in 2007. Total economic losses leapt to record $232 billion from $82 billion in 2007. Munich Re said the year was marked by high losses from weather-related natural disasters, continuing a long-term trend. Munich Re said in a study with the UN Environment Programme that weather-related disasters seemed to be on the rise, **in line with forecasts by the UN Climate Panel.** "Since the 1980s, earthquakes have risen by around 50 percent but weather-related hazards such as major floods have increased by as much as 350 percent and those from wind storms have doubled," the report said. - http://www.huffingtonpost.com/2008/12/29/natural-disasters-cost-in_n_154004.html

"During the 2000 to 2009 period, there were 385 disasters, an increase of 233 percent since 1980 to 1989, and of 67 percent since 1990 to 1999, according to CRED data." Though earthquakes made up 60 percent of natural disasters from 2000 to 2009, climate-related events, such as droughts, storms and floods, have made up the majority of disasters overall, increasing tenfold since data was first collected in 1950. "Have climate-related disasters increased? The answer is yes," Director Debarati Guha-Sapir, Center for Research on the Epidemiology of Disasters - The News - Natural Disasters, January 28, 2010

"**Climate change** is also being blamed for a long-term trend toward severe flooding, according to a report issued today by the Interagency Climate Change Task Force."

"In between, tornados, floods and massive snowstorms have caused the loss of lives and massive property damage. According to scientists who study **global warming,** these are just a few of the natural disasters predicted in the years ahead." – Peterson Thomson Marketwire - March 3, 2010

"but one thing is certain, the number of natural disasters in the past 20 years outweighs any other written recorded history." - Sheryl Young - March 8, 2010

"Munich Re, the world's largest reinsurer, has said losses from natural catastrophes caused by **climate change** will rise." - Jamie McGee - Mar 16, 2010

The increasing intensity of natural disasters

Over the past decade we have seen an increase in the intensity of natural disasters worldwide. In the recent past, we have had natural disasters created by the tsunami in Asia, the earthquake in Pakistan, and the hurricane Katrina in the US. These caused unprecedented devastation and great loss of life which have been etched in our minds due to the magnitude of the devastation.

Ref: http://www.unisdr.org/disaster-statistics/occurrence-trends-century.htm

Number of natural disasters registered in EMDAT
Across the years 1900-2005

Source of data: EM-DAT : The OFDA/CRED International Disaster Database.
Http: //www.em-dat.net, UCL - Brussels, Belgium

According to an article published by lifesciences.com, in the year 1980 only 100 such disasters were reported. However that number has risen to over 300 a year since 2000. The climate story of the decade is that the 2000s are on track to be nearly 0.2°C warmer than the 1990s. And that temperature jump is especially worrisome since the 1990s were only 0.14°C warmer than the 1980s. The first decade of this century is "by far" the warmest since instrumental records began, say the UK Met Office and World Meteorological Organization. (Source: The BBC and Climateprogress.org)

By Rowland Croucher and others ☐ February 5, 2011
In the first update to his *2008 Climate Change Review* on Thursday, Professor Garnaut was clear about the link between recent extreme weather events – from Black Saturday to cyclone Yasi – and global warming. What's more, he flagged that since his 2008 review, the science has only become more alarming.

With warming now at less than 1 degree above pre-industrial levels, and with the sort of emissions growth coming from the industrialisation of China, India, Indonesia and other developing countries, "if we are seeing an intensification of extreme weather events now ... you ain't seen nothing yet".

While much policy discussion is about limiting warming to 2 degrees, Spratt writes, the scientists are telling us it should be kept to under 1 degree, and the planet is actually heading towards 4 degrees due to "chronic political failure".

Dear Madam/Sir,

I remember how strange it was in recent years to notice the New South Wales bushfires were starting to happen every year instead of once in a blue moon. Furthermore, you can remember in just the last few years Australia witnessed that unprecedented Dust Storm that swept across the majority of our continent, the unprecedented extremity of the Victorian Bushfires, the unprecedented floods in Queensland accompanied by the subsequent largest ever Cyclone. As I was growing up I could only ever remember such events happening once every ten or more years with nowhere near the intensity of what has recently occurred.

We are now starting to see what the scientists predicted only a few years ago, we are starting to pay the economic price for climate change and the majority of **Scientists** along with the **Insurance Companies** are predicting it to continue and worsen.

Surely now, how can we question whether or not to implement the Carbon Tax; when this is unfolding in front of our very eyes and when the rest of the world, to a large degree, is adopting the same or similar measures. China has just announced it will reduce its Carbon output by 17% in the next five years.

I urge you, please support the Carbon Tax and sleep with a clear conscience.

Yours sincerely,

Lincoln Coull

Concerned Australian

It was after having sent those emails that I believe a definite, more serious resurgence of a collective global Climate Change movement grew.

As I had collected a lot of media email addresses in California in the US, I started to notice a new genre of movies being released with some sought of Climate, Save the Earth related type of message. Movies like The Day the Earth Stood Still, Avatar, Geostorm and Tomorrowland were encouraging to watch.

With the rise of people through this era like award winning author Naomi Klein, activists Greta Thunberg and Bill McKibben, US Senator Bernie Sanders and US Minister Alexandria Ocasio-Cortez, and organizations like The Guardian and Extinction Rebellion. Also, Leonardo DiCaprio's poignant documentaries, The 11th Hour and Before the Flood and the most significant efforts of Mr. Gore and President Obama. In addition to the constantly occurring extreme natural disasters, the Climate Crisis increasingly gained more of the real world attention needed and has become if not the most thought about issue. The seriousness of Climate Change continues to reach the global vast majority more so now than ever before.

Hence in time, all this contributed to the landmark Climate Change Agreement induced at the COP 21 meeting in Paris, December 2015. This historic meeting culminated with 195 nations formally agreeing to keep the global temperature rise to below 2 degrees Celsius for the remainder of this century and to further implement measures to limit the increase to 1.5 degrees Celsius above pre industrial levels.[1]

I have to say, witnessing that had me raising my arms with sheer joy and elation.

What a monumental achievement for humankind and Mother Earth.

Looking out the back screen door, I saw a family of six magpies frolicking on the grass around my back yard. It was very rare to see so many together there. Instead of venturing to lay in my hammock or sit in one of my outdoor chairs under the pergola, I chose to walk toward them and sit on the edge of the pavers meeting the lawn. The magpies, the mum and dad and the four younger siblings didn't move away as they usually did, they seemed to readily accept me into their space, not a metre from them. Falling short of almost feeding them from my hand, this interaction was so life affirming.

Then the following summer, while fishing on the beach with Dad on a clear blue sun shining day, we were sitting in our deck chairs and there was an unusual stillness. Looking out across the water in the distance, I noticed a distinct solitary cloud in the shape of a Crown floating towards us. Dad and I both acknowledged how strange it was; I took a couple of photos.

The Commission

A couple of years later, things were progressively going pear shaped on a few levels. I began to be regularly psychologically and materially attacked from multiple media sources, tv, newspaper, radio, online, at work and at home. Admittedly, most of the media assault was done in an ambiguous, innuendo type of way, but then I'd wake up in the morning to find someone had messed up my back yard in the night or I'd come home after work to find the ceiling fan running in the middle of winter and my car having been broken into; nothing stolen just bits chucked about everywhere. I held silent for many, many months. Then, when it became too much to bear, on a particular Saturday, I put something on the face of my computer screen while on the internet. It was 'No more War' and 'No more Child Sex Abuse'. I left it on the screen all afternoon while still on the internet. Surprisingly, straight after doing that, when I opened the front door to go to the beach, I looked up and saw a massive Crown cloud covering the whole sky.

To my astonishment, the next day, the Prime Minister declared the start of a Child Sexual Abuse Commission. Unexpectedly, the following Friday (11/01/2013), I was **Bcc'd** a joint email from the Prime Minister, Attorney-General and the Minister for Families, Community Services and Indigenous Affairs that was addressed to the Royal Commission Secretariat announcing "the Royal Commission into the institutional responses to child sexual abuse".

Julia Gillard
Prime Minister of Australia

Nicola Roxon
Attorney-General

Jenny Macklin
Minister for Families, Community Services and Indigenous Affairs

Good afternoon

We are very pleased to announce that on our advice the Governor-General has now appointed six Commissioners for the Royal Commission into institutional responses to child sexual abuse. We have also today released the Terms of Reference that will guide their inquiry.

Later on, I found myself sensing a real profoundness in contemplating how I had significantly contributed to and or influenced the three most disturbing things that were constantly on my mind; Saving Australian Culture, Climate Change and Child Sex Abuse.

Nonchalant Heroes

In December 2016, happy from being given half the day off work, waiting at the bus stop, a well-dressed old Indian gentleman started up a conversation. We discussed how Adelaide was such a great place to live in comparison to the other Australian capital cities. He told me that he'd traveled the world as a jazz musician playing the guitar and that he'd migrated from India with his family in 1973.

As it was crowded on the bus, we sat together continuing our conversation. He told me that back in India he had managed a five-star hotel for years and had met quite a few of the cricket greats and other celebrities while working there. Aside from telling a funny story about a South African cricketer and commentator, the first celebrity he mentioned meeting was Sir Edmund Hillary. After having played a few mountain songs for him, Hillary approached and thanked him and they continued chatting from there. I mentioned he seemed to be a gentleman and he agreed, but then said he was seen as a hero to the Sherpas over there as he had returned many times to help them and their community. Then 83, the Indian gentleman mentioned he had met quite a few heroes in his time; some big ones who did major things to help others. He said many of them, you wouldn't think were a hero; wouldn't tell you or mention it. He added, but then there are a lot of ordinary people who are heroes also, like your neighbor, helping out and doing all sorts of stuff that you wouldn't know. I acknowledged that, saying they are nonchalant and everyone should contribute in some way and help others.

As I rose to get off the bus, he stood and shook my hand wishing me all the best and I told him I enjoyed our chat and wished him a happy new year.

Off the bus walking to the train station, rehashing our conversation, I started to humbly praise God.

A New York Treat

I flew to New York in early 2016.

On visiting Madame Tussauds, I remember turning right to walk through the door and unexpectantly came face to face with President Lincoln sitting behind his desk. As I passed him and proceeded into the room, to my amazement, I found myself gazing upon Mahatma Ghandi, Nelson Mandela, the Deli Lama and the current Pope Francis. Looking further to the right was President Obama and his wife Michelle.

After the picture taking, in the middle of the room viewing them, I realized I was amongst my all time, most highly respected humans. I stood in absolute awe, which was followed by a heightened sense of acknowledgement.

Respectfully, shortly after, I regretfully noticed two others, Martin Luther King and Mother Teresa, weren't there.

Nevertheless, later, in reflection, I couldn't help but feel very thankful for the experience.

Although I've never needed to look up to anyone, I always held the afore mentioned, in the highest regard and they certainly influenced me. Whatever their quest, these are the great humanitarians; the great examples.

Relatively, in thinking of all these influential leaders, it reminds me of what I learned of Confucianism at University. The people living in a village following Confucianism, would only choose the leader that had showed the highest ethical values and had proven practical examples of virtuous conduct. They then would have the peace of mind knowing that person will do the right thing by them.

The Flag

Interestingly, somehow I've come to see my path with more clarity recently. I realize that all I've experienced in the past, the challenges and the triumphs, and this might sound a bit cliché, I've had to learn from and that has contributed to my evolving to this point. Evolving to become a better human being; perhaps for someone most worthy.

When I planted that flag on top of the mountain, yes I was fulfilling what I perceived to be a necessary, sincere quest. I started this because I was and still am very passionate about Saving the Earth and Humankind.

However, while making my flag…. I realized it may also have another purpose….

I intuitively sensed I would find true love somewhere along this endeavour. I always wanted nothing less, didn't want to settle for second best.

Madam

In September 2019 while laying on the lounge watching TV, I received a mysterious text message on my mobile phone thanking me for asking them about how to contact one of Hollywood's most influential A1 stars. This was somewhat startling as I had never put anything forward online or otherwise asking this.

I proceeded to start writing letters every two weeks to her. As there started off being eight addresses, I had to hand write eight copies each time and send them off in the post. As time went by, the addresses gradually reduced to my writing just the same letter just twice. Over the following two and half years, I wrote her a total of 48 letters.

I started by stating "I'm coming from the heart as that's the only way this will work".

With her tireless Humanitarian work and knowledge of my quest from my emails to the media, it seemed fitting that we felt an affiliation, a mutual respect and acknowledgement. Hence, I preferred to be with a woman who shared similar traits and knows most all of my story; as she did.

During that time, I also found myself wondering and somewhat taken back by our interaction at this time in the world. How two peoples' very positive, altruistic efforts possibly leading up to a meeting amongst these current global events; when the world needs more love and care, more now than ever before.

I prefer to be with my soul mate and I wanted to see if we had that. I sensed it was somehow destined that our souls were meant to meet in this lifetime. I seriously thought we had a chance of sharing a real bliss and contentedness. As our selfless efforts have been well above the norm, I felt more than ever, that she and I were on some sought of similar ascending journey.

One night in bed just before falling asleep, she appeared to me for about two seconds. I saw her clear smiling face and hair, wearing a white robe and she was glowing with light maroon and blue colours shining out around her.... I've come to learn and accept that it was very likely a 5th dimensional experience. I've never had, felt or experienced anything like that before.

Later, I dreamt we were together and she was driving with a child in the back seat. It was a sunny day and her hair was shorter, just above shoulder length and we were going back to her place. Then later, in mingling with others, I was impressed with her confidence in expressing her knowledge to them. After that I think, we were at my place and then as the vision faded off, the dream ended showing the word "agreement".

About four days after that, I watched a tarot reading on YouTube and the presenter said there is a divine intervention which brings a defining moment in my life. She said me and that other particular sign had a past life connection and we made an "agreement" to meet in this lifetime for a specific reason. And that we are awakening together in this life at the same time and are about to go through a transition together. She also said Arch Angel Michael is protecting this union.

The tarot reading verified my dream.

I consider myself to be a spiritual being, I believe in God, but I never expected to experience this kind of spiritual enlightenment or insight.

Although there were a series of abstract references toward me in her some of her movies and online, there was never a direct response. She didn't meet me in Fiji and I didn't meet her in Asia. I called it off a couple of years ago; the meeting never transpired.

Idealism

For over thirty years now I've been wearing my sterling silver Peace bracelet. When I was about 18, I had a constant vision of it in my head for about four months before finally going to a jewelry shop, drawing a picture of it and asking to have it made. After picking it up, I went to a church and asked a Priest to bless it. I had always preferred Peace over most everything else.

Many years ago I watched a documentary whereby a group of anthropologists discovered the oldest human civilization. I think it was in Southern America. The anthropologists were puzzled as to why they couldn't find any fences or walls around the city. They went about finding and analyzing all the artifacts. They found fishnets made from the cotton they grew, even though they lived hundreds of miles inland. They learned that the people traded the fishnets with the tribes that lived on the distant shores, for fish and other goods. The anthropologists concluded that they had built up and established an extensive trading network that covered the whole continent and further. Thus, the whole site proved the first human civilization was born out of Peace and lived that way for a thousand years without conflict and trade enhanced and encouraged it. I've always believed we could end up living globally that way again. I know many would argue this to be an unrealistic or naive perspective. But President Clinton knew this. He invested a lot of his time in promoting his World Trade Organization agenda, as he saw it as a vehicle to establish more peace and stability in the world.

I think it's relative to mention another documentary I watched decades ago that conveyed what a leading Anthropologist had learned in his years of study both in literature and in the field. He studied numerous different peoples around the globe, from the Wodaabe tribe in Africa, to a Himalayan goat tribe, to modern day New York culture and many others.

He came to the conclusion that of all the different cultures he studied, they all shared the same fundamental principles: 1/. Family 2/. The Arts, Music and Dance and 3/. Some sought of Worship and or Religion. His further conclusion was that if you as an individual have all three of these things in your life, then you should be somewhat fulfilled and balanced. However, I would go a step further, in that for those living in our modern economic time, a fourth element is necessary. One should also have a reasonable amount of money to sustain a somewhat comfortable life.

The **Human Rights Declaration**; I remember how uplifted I was learning of it at university and how it was formed to protect people and their fundamental Rights. I still have a hard copy of it today.

Relatively, I was glad to read Greta Thunberg, with others, was successful with her Human Rights complaint. The Human Rights Panel concluding that the big fossil fuel companies can or could be sued for being responsible for their carbon emissions.

But with all this that I've mentioned in this Idealism chapter, you could well ask why, what's the relevance?

When I look back on all of my readings and education, I came to wonder, are these principles of Peace, Freedom, Fulfillment and Human Rights just words to be merely considered and or ignored. You think further of other words such as Respect, Virtue and Morals and how they came into existence. I suspect they manifested over the eons of time essentially from human experience and intuition. Moreover, I've come to believe that God instilled them in us. Then one asks, why?

Because they represent ideals to help steer us in the right direction, towards living a more righteous life. Over time, I've come to appreciate the value in that.

Spiritually

As hidden as some of this was, there are many eyes on me and as well as Gods' eye who sees everything.

Thanks largely to my mothers' devoutly religious influence - and though in my younger years I never thought I would - over the last two years or so, I've found myself, meditating, praying and cleansing my soul. Connecting with God and meeting the refining demands on my path.

After having planted my Save Mother Earth flag on the mountain and sent out all those thousands of emails around the globe, whilst in a church one sunny day during solitary prayer, God communicated to me the word "Protector" and within the next millisecond said I'm "Protected".

I'm eternally humbled as God continues to communicate to me in other ineffable ways.

After Pope Francis released his Encyclical Letter, LAUDATO SI', On Care for Our Common Home in 2015, I spent one whole afternoon in my hammock underneath my backyard pergola determined to read all of it. After finishing the last sentence, I looked up to a clear blue sky on my left and saw a solely cloud formation in the shape of the word "Hi" and quietly, humbly acknowledged his presence.

Still, each morning just before getting out of bed, I make an effort to thank God for this day.

And later, on the way to work I pray and thank God for my life and his daily blessings.

How we got here

40 years of the Republican dog eat dog ethos

Although there are traces of Capitalism in Roman times, modern Capitalism really started in Medieval Europe from Feudalism with the interaction of landlords and agricultural producers. However, the expansion of steam and coal power in England in the early 1800's saw Capitalism really take off and in turn, with the rise of machines and factories, it led to the Industrial Revolution. As agricultural efficiency and housing conditions vastly improved, this significantly contributed to a global population explosion in what was then viewed a limitless world. Everyone was blissfully unaware of the vast volumes of carbon dioxide and other greenhouses gases that were being emitted into the atmosphere as the Human race significantly advanced.

It's worth mentioning that to its credit, Capitalism has also inspired and provided some amazing innovations like the car, train, plane, telephone, television, fridge, computers and encouraged all the medical and technical luxuries we enjoy today.

Fast forward to contemporary times and the main sources of Climate Change today are from the major Capitalistic sectors, primarily the burning of fossil fuels, then industry, deforestation, transportation and the farming of livestock.

Essentially, it is now evident and scientifically proven that humans, indeed our actions, are the direct, fundamental cause of Climate Change.

Most all the studies and graphs show a sharp incline in Greenhouse gas emissions starting in the 1950's that coincides with sharp inclines in Global Warming surface temperature and the Global Ocean temperature starting in the same decade. As the following graphs show:

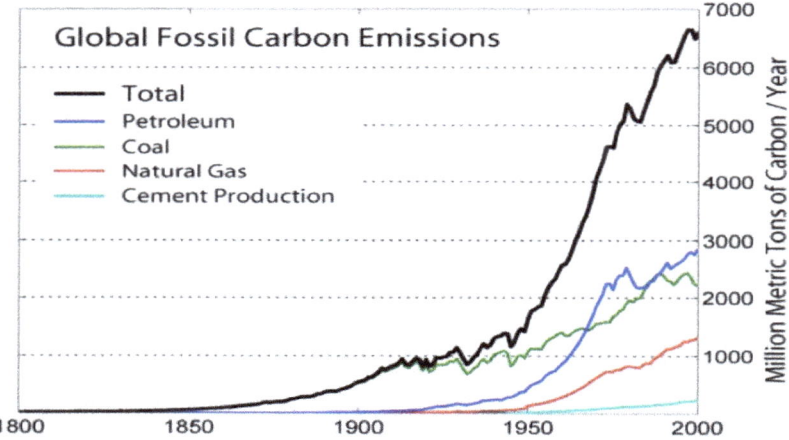

Sourced from Wikimedia Commons, the free media repository - Global annual fossil fuel carbon dioxide emissions, in million metric tons of carbon, as reported by the Carbon Dioxide Information Analysis Center

Global ocean heat content, 1950-2021

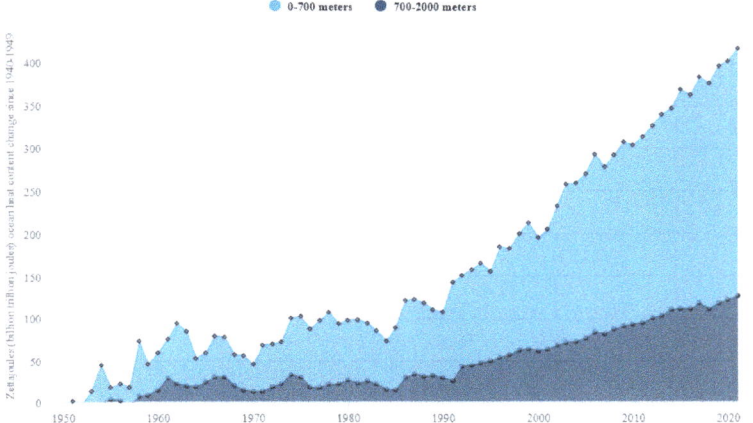

Annual global ocean heat content (in zettajoules – billion trillion joules, or 10^21 joules) for the 0-700 metre and 700-2000 metre layers. Data from Cheng et al 2021. Chart by Carbon Brief using Highcharts.

Global Warming – Earth is Boiling

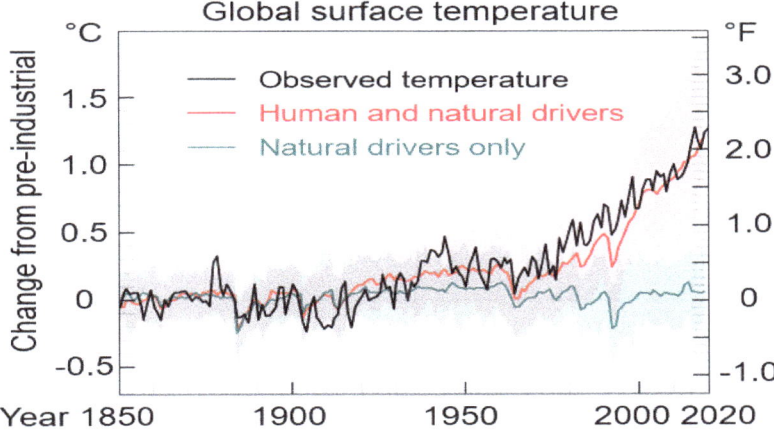

Sourced from Wikipedia - Observed temperature from NASA vs the 1850–1900 average used by the IPCC as a pre-industrial baseline. The primary driver for increased global temperatures in the industrial era is human activity, with natural forces adding variability

In the mid-70s, there was so much promise. America was progressive under President Jimmy Carter who had passed "14 major pieces of Environmental legislation", including the Clean Air Act, the Clean Water act and funding for alternate energy. "He commissioned the Global 2000 Report" that analysed environmental impacts and sustainable development. Under his Presidency, his "White House Council on Environmental Quality (CEQ) issued three reports", with the last one focusing on the long-term effects of "carbon dioxide pollution".[1] He even had Solar Panels installed on the rooftop of the White House.

But then, the perspective began to change as was revealed in the release of Nathaniel Rich's article "Losing Earth: The Decade We Almost Stopped Climate Change" in August 2018.[2]

It showed that in 1974 the CIA, President Jimmy Carter and the worlds' biggest fossil fuel company, Exxon Mobil were all well aware of the increasing Carbon Emissions and resulting Climate Change problem.

The following caption shows that in July 1977 a memo was presented to President Carter from his chief science adviser and director of the Office of Science and Technology Policy, Frank Press, warning of a future climate catastrophe due to the CO2 emissions from fossil fuel use:

THE PRESIDENT SEEN.

EXECUTIVE OFFICE OF THE PRESIDENT
OFFICE OF SCIENCE AND TECHNOLOGY POLICY
WASHINGTON, D.C. 20500

July 7, 1977

MEMORANDUM TO   THE PRESIDENT

From: Frank Press

Subject: Release of Fossil CO_2 and the Possibility of a Catastrophic Climate Change

Fossil fuel combustion has increased at an exponential rate over the last 100 years. As a result, the atmospheric concentration of CO_2 is now 12 percent above the pre-industrial revolution level and may grow to 1.5 to 2.0 times that level within 60 years. Because of the "greenhouse effect" of atmospheric CO_2 the increased concentration will induce a global climatic warming of anywhere from 0.5° to 5°C. To place this in perspective, a ΔT of 5°C would exceed in 60 years the normal temperature swing between an ice age and a warm period which takes place over tens of thousands of years.

The potential effect on the environment of a climatic fluctuation of such rapidity could be catastrophic and calls for an impact assessment of unprecedented importance and difficulty. A rapid climatic change may result in large scale crop failures at a time when an increased world population taxes agriculture to the limits of productivity. The urgency of the problem derives from our inability to shift rapidly to non-fossil fuel sources once the climatic effects become evident not long after the year 2000; the situation could grow out of control before alternate energy sources and other remedial actions become effective. Natural dissipation of CO_2 would not occur for a millenium after fossil fuel combustion was markedly reduced.

As you know this is not a new issue. What is new is the growing weight of scientific support which raises the CO_2-climate impact from speculation to a serious hypothesis worthy of a response that is neither complacent nor panicky. The authoratative National Academy of Sciences has just alerted us that it will issue a public statement along these lines in a few weeks.

The present state of knowledge does not justify emergency action to limit the consumption of fossil fuels in the near term. However, I believe that we must now take the potential CO_2 hazard into account in developing our long-term energy stragegy. Beyond conservation, we must be prepared to exploit nuclear energy more fully. As insurance against over-reliance on a nuclear energy economy, we should emphasize targeted basic research which could lead to breakthroughs for solar electric, biomass conversion or other renewable energy sources. I am already working with OMB and other Federal agencies on a national climate research program which would lead to a better assessment of the CO_2 hazard. If you agree, I will work with OMB, ERDA, FEA, and NSF on alternate strategies for R&D, responsive to a possible CO_2 hazard.

Electrostatic Copy Made
for Preservation Purposes

However, the above memo was also presented to the President with an advisory note from the then American first secretary of energy, "James "Jim" Schlesinger":

```
My view is that the policy implications of this issue are
still too uncertain to warrant Presidential involvement
and policy initiatives.
```
[4]

Further to that point, in 1983, it was presidential advisor William Nierenberg, who initially acknowledged that immediate action was warranted, then hypocritically decided to advise, that it was not necessary to act then and leave it to later generations to solve, as they would be more technically advanced to deal with it. Thus, "George Keyworth II," President "Reagan's science adviser, used Nierenberg's" view and "warned against taking any "near-term corrective action" on global warming. Just in case it wasn't clear, Keyworth added, "there are no actions recommended other than continued research."[5]

Similarly, also in July 1977, Exxon Mobil's senior scientist James Black advised their management committee that "mankind is influencing the global climate is through carbon dioxide release from the burning of fossil fuels."[6] Instead of acting responsibly with this knowledge, Exxon subsequently "became a leader in campaigns of confusion" on climate change. It also "helped to prevent the US," "China and India, from signing" the 1998 "Kyoto Protocol to control greenhouse gases."[7]

Thus, it would be apt and plausible to postulate that if the American government acted in favour of protecting the Environment back then, by implementing legislation to reduce Carbon Emissions and those laws remained, the rest of the world's Governments would have followed and we would not be dealing with this Climate Crisis now.

Mindful that Government was initially formed to serve and protect the people;

the last 40 years of neo liberalism – US Republican dominated politics - has proven only to drastically hinder action on Climate Change and further enrich the wealthy to the real detriment of the majority of the human population, other species and the Earth.

President Reagan quoted (1980): "we're going to turn the bull loose." And Britain followed. They gambled that an unregulated free market would expand capitalism and therein increase employment, spending and general prosperity…. the Opposite transpired, eventually resulting in the 2008 GFC.

It encouraged an unequalled selfishness that catapulted the rich to the super rich. The global middle class was significantly decreased with a vast majority having a reduced quality of living or loss of their lively hood. Massive CO_2 pollution also ensued and further enhanced the Climate Catastrophe.

Hence, the Neo Liberalism theory and application conclusively failed…

The Contemporary Anthropocene era

Thinking of those worst hotspots like Yemen, Syria, Myanmar and others. The people in those areas seriously suffering; the hungry, the homeless, the sick and the victimization of women and children. As bad as it all is, how bad would it be without organizations like Amnesty, Unicef, Red Cross, the UN and the countless NGO's. I take my hat off to all of them for continually tackling these immense tasks and atrocities.

All this outside of Climate Change, yet as you know, in recent years Climate Change has been contributing to these conflicts and suffering and will continue to do so on an ever-increasing scale in the future.

It's hard to ignore what's unfolding around the globe.

In the northern hemisphere 2022 summer, we witnessed unprecedented heat waves that led to many countries in Europe; Italy, Germany, France, Greece, Portugal, Spain and the US all suffering from wildfires. Most all of Europe was in extreme drought, experts predicting the worst in 500 years. Frances' drought, the longest ever recorded. While Hong Kong China recorded it hottest ever week. This reminds me of the recent, never seen before, fires in constantly cold, wet states like Tasmania and Alaska, caused by unprecedented massive lighting storms. The astonishing and unprecedented nationwide Australian bushfires of 2020. All this resulting from lengthy periods of no rainfall and record high temperatures. Then, the more recent devastating record breaking flooding along Australia's east coast. Parts of the state of New South Wales experienced their fourth consecutive flood inside six months; each one higher that the last. Unheard of.

More importantly, it's evident now that the yearly examples of extreme Climate Change events we have been witnessing, e.g. record storms and floods, giant wildfires all happening around the world, **have stepped up a notch.**

They've gone from affecting the odd area to affecting whole countries and whole regions.

These events and the near collapse of the Gulf Stream, is further evidence that we are indeed experiencing real, direct, adverse effects of Climate Change right now and have been for most of this century.

Unfortunately, I fear these extreme Climate Change events are just the initial stages of what's to come if we don't drastically and immediately turn things around.

Furthermore, with the George Floyd murder, Covid 19, plastic smothering our oceans, the ongoing Climate Change Disasters, Putins' insane war, the real threat to global food security, logistic issues and inflation, it has all added to the masses' general disappointment and frustration toward Government inaction.

And while other resulting adversities like the increase in domestic violence, displacement and homelessness get some air, they continue to go largely unnoticed and unaddressed.

For too long, Governments have been avoiding their fundamental principles of acting in service and protection for the People and the Environment and abhorrently bowing down to aiding the global fossil fuel corporates.

Conversely, I'm also convinced that all of these recent events have also given rise to every person re evaluating what really matters.

Thus, Governments need to listen, change and act for the People and the Environment first, now more than ever before. Relatively, if the corporate companies of the world, especially the biggest ones, were heavily regulated - and they should be - they would still make their millions and billions. Not to mention their not paying tax or a fairer amount of tax. For if the corporates did pay the appropriate tax – currently, there is talk of inducing a 15% global corporate tax (the US recently legislated it there) – and Governments stopped paying tax payers money, trillions of dollars in subsidies to the fossil fuel industry, then they could allocate more funding to the Green Renewable Energy Sector and Green infrastructure and therein produce millions of new jobs around the world.

We can, need and must wean ourselves off all fossil fuel use forever.

Positive Reflections and Solutions

What we are doing

There are examples that replenish my faith in the human spirit.

It warmed me to see an overall united effort of the global family working together to have seemingly beaten the Covid virus now.

However, reflecting on the Climate Change situation, I remain the glass half full kinda guy, as we have the Science, Technology, Manufacturing base and the Funds, I know, as we worked together to beat the virus, we can rise together to overcome this, the biggest challenge in the history of our species.

Relatively, it was very encouraging to view the recent Australian 60 Minutes segment (01/05/2022) showing Australian mining billionaire Andrew Forrest and Australian Tech Billionaire Mike Cannon-Brookes both claiming we have the technology now to service the entire worlds' energy by renewable sources and it is economically viable and profitable.[1]

Andrew Forrest quoted "we need to move the world on from fossil fuels" and intends to show the world that it can be done while incorporating the lowering of operating costs, increasing production and increasing profits.

His company ran a mass scientific analysis that showed there is enough green energy to supply the whole world without using any fossil fuel ever again. He further claimed that "a small percentage of Australia can completely power the world".

His Fortescue Future Industries Company has spent the last 10 years developing green Hydrogen energy to power all of his operations and to also power the whole of Australia.

As "Hydrogen is cheap and has huge earning capacity in exports" he has a strategy in place now to produce 15 million tonnes of Hydrogen a year by 2030. His has built a hydrogen fuelled truck, train and ship. He has further funded $3 billion dollars of his own money to, with the Queensland Government, build the world's largest electrolyser facility in Gladston that will produce green Hydrogen.

His view of big industry leading the charge, has been long overdue and no one has wanted to go first. He points out that as there is a lack of Government policy that should lead business and industry, Business then, needs to show that major change is both commercial and responsible.

Andrew boldly and correctly states that the previous government stood in the way of "the market, climate science and that we must stop this world from cooking".

Additionally, Australian Tech Billionaire Mike Cannon-Brookes, says with our abundant natural resources and geographical position, we have the talent and finances to service 3 billion consumers to the north. He says "we can be the largest exporter of energy; the renewable energy Superpower". He is currently still developing the "Sun Cable Project" to sell Singapore our renewable solar energy and deliver 20% of their energy needs.

Currently, 30% of all Australia's energy comes from Green Renewable energy and Mike says that can be increased to 100% and more. He wants to change AGL Coal fired Power Stations to Green Hydrogen and has been knocked back; but he has since bought 11% shares of that company. And if successful in his pursuit, it would have a domino effect around the world.

Europe and other countries are making real efforts to also combat it. Countries like Sweden, Costa Rica, Nicaragua, Scotland, Germany, Uruguay, Denmark, Morocco, Kenya and South Australia have all achieved impressive Renewable Energy targets.[2] Incidentally, having installed the worlds' first biggest battery in 2017, in just 15 years, South Australia smashed its initial Renewable target, having transformed its energy system from 1% to over 60% and its Wind and Solar power has met 100% or more of local demand on numerous occasions.[3] It intends to achieve 100% continual Wind and Solar energy by 2030 and further intends to generate 500% of its energy needs from renewable sources by 2050, with the surplus to be exported nationally and internationally. However, while most all countries have pledged Net Zero emissions by around 2050, more importantly, it's countries like France, Greenland, Denmark, Belize, Spain and Ireland that have legislated the ban on extraction and production of fossil fuels or new exploration of them.[4]

Damon Gameau's documentary "2040" shows great examples of how in Bangladesh they are implementing Solar home systems that via a solar panel, battery and a Soleshare box, each home gets the electricity they need, but they also connect and share energy amongst each household in their village creating a micro grid.[5] The micro grid connects from village to village and spreads across the country. Instead of purchasing energy, as the network grows, they sell electricity and the money is kept locally which in turn, grows small local businesses and benefits their village's economy. There proves to be many other onflow benefits, like being able to read and educate themselves at night with the light bulbs powered by this newly attained electricity.

It also evinces the ambitious efforts of Paul Hawkens' "Project Drawdown" that aims to reverse global warming via regenerative development. The project influences the change of food and agriculture practices to stop emitting greenhouse gases and sequester them. They encourage the development of Agro Forestry whereby highly diversified food systems grow multiple food crops in the same plot of land. E.g. Pawpaw, Bananas, Coffee, Avocados' and Veges. It also highlights Dr. Brian Von Herzen's work on the very promising Marine Permaculture that has multiple benefits like Seaweed production that can produce food, animal feed, fiber, fertilizer, biofuel and transforms the quality of the seawater. They claim it could potentially feed 10 billion people.

It also raises the serious point that if more girls around the world were to complete their education, there would be less child births and inevitably less demand on global resources.

Finally, it sheds light on the great example in Sweden's city of Stockholm, whereby they collect the resident's food waste, extract methane from it to then power the local government vehicles and the leftovers are passed onto the farmers to use as fertilizer.

As Sir David Attenborough tells us, "many countries are planting trees by hand" and they're also using technologies like deploying drones to drop seeds in remote areas.[6] Costa Rica has concluded that to have a healthy economy and society, you must have a healthy Ecosystem. The Government rewards landowners for protecting and restoring forests. It's the one of the very few countries to have successfully reversed Deforestation and has recently been awarded the "United Nations Environment Programme's Champions of the Earth award.[7]

The following cities, London, Aarau, Bolzano, Milan, Oslo, Hamburg, Phileas Cologne, Whistler, Perth and Beijing have all trialed Hydrogen fuel cell busses.[8] We have synthetic oils for our cars now and there are cars and trucks that are running on electric battery power. We have harnessed the power of the Wind and the Sun on national scales. We have built enormous batteries to compensate for the down time.

In July 2022, Microsoft successfully built and tested its own Hydrogen fuel cell.[9]

As an emissions' free, clean, abundant energy source, everyone has acknowledged its potential and many countries have entered the Hydrogen race. They have or are setting aggressive targets and strategies in this space. Hydrogen fuel will soon deliver globally.[10]

Furthermore, proudly, on 21st of August 2022, South Australian Premier Peter Malinauskas announced his state government is about to build the world's biggest Hydrogen plant to be operational by the end of 2025.

There is no excuse for using fossil fuels anymore; no more digging or drilling.

While COP 21 in Paris was a great step forward, could it be that we may see major sweeping change in our lifetime?

It so desperately needs to happen as we are literally running out of time now.

We may have turned the corner?

Although as you would know, there is still a mammoth task ahead of us, a huge amount yet to be done in tackling Climate Change, it's possible we may have turned the corner.

I was aware that the recent Australian Election was to be on the 21st of May 2022.

Due to the afore mentioned Climate catastrophes that many Australians were still suffering from, I inherently saw an opportunity to bring the Climate Change Crisis once again to the forefront and every level of the Australian consciousness. I thought it was a good time to encourage a new, hopefully, Labor Government to get on the front foot and capitalize on the countries' growing Climate Change psyche. It was a chance to also encourage an end to the decade long climate wars in our country and implement a new Climate progressive era.

Just two weeks before the election on the 7th of May 2022, that Saturday was a rainy day, so it was easier for me to commit the whole day to sending emails out across the nation and I was determined to also email every Australian political representative, every Minister and Senator. I proceeded to send the following letter:

WE MAY HAVE TURNED THE CORNER

This article emailed to most every Australian Newspaper, TV and Radio Stations and the majority of Ministers/Senators in the Labor Party, The Greens, Liberal Party and some Independents. 07/05/2022

Dear Senator/Minister,

Please allow me to convey my thoughts in the lead up to this crucial election.

Mindful of the past 15 years of Government inaction on Climate Change, it was very encouraging to view the recent 60 minutes segment (01/05/2022) showing Australian mining billionaire Andrew Forrest and Australian Tech Billionaire Mike Cannon-Brookes both claiming we have the technology now to service the entire worlds' energy by renewable sources and it is economically viable and profitable.

Andrew Forrest quoted "we need to move the world on from fossil fuels" and intends to show the world that it can be done while incorporating the lowering of operating costs, increasing production and increasing profits.

His company ran a mass scientific analysis that showed there is enough green energy to supply the whole world without using any fossil fuel ever again. He further claimed that "a small percentage of Australia can completely power the world".

His Fortescue Future Industries Company has spent the last 10 years developing green Hydrogen energy to power all of his operations and to also power the whole of Australia.

As "Hydrogen is cheap and has huge earning capacity in exports" he has this strategy in place now to produce 15 million tonnes of Hydrogen a year by 2030. His has built a hydrogen fuelled truck, train and ship. He has further funded $3 billion dollars of his own money to, with the Queensland Government, build the world's largest electrolylizer facility in Gladston that will produce green Hydrogen.

His view is big industry leading the charge has been long overdue and no one has wanted to go first. He points out that as there is a lack of Government policy that should lead business and industry, Business then, needs to show that major change is both commercial and responsible.

Andrew boldly and correctly states that the current government is standing in the way of "the market, climate science and that we must stop this world from cooking".

Additionally, Australian Tech Billionaire Mike Cannon-Brookes, says with our abundant natural resources and geographical position, we have the talent and finances to service 3 billion consumers to the north. He says "we can be the largest exporter of energy; the renewable energy Superpower". He and Andrew are currently developing the "Sun Cable Project" to sell Singapore our renewable solar energy and deliver 20% of their energy needs.

Currently, 30% of all Australia's energy comes from Green Renewable energy and Mike says that can be increased to 100% and more. He wants to change AGL Coal fired Power Stations to Green Hydrogen and has been knocked back; but he has since bought 11% shares of that company. And if successful in his pursuit, it would have a domino effect around the world.

Mindful of the obvious yearly examples of extreme climate change events we are witnessing now, e.g. the rising seas, record storms, giant wildfires and floods all happening around the world constantly and the latest IPCC Report published on 9th of August 2021 confirming "Human-induced climate change is already affecting many weather and climate extremes in every region across the globe. Evidence of observed changes in extremes such as heatwaves, heavy precipitation, droughts, and tropical cyclones, and, in particular, their attribution to human influence, has strengthened since the Fifth Assessment Report".

I put it to you, will you act responsibly now:

- Will you implement legislation to cease the use of all fossil fuels?
- Will you implement legislation to cease the use of all tax payers' fossil fuel subsidies to the fossil fuel industry and re direct them to the Renewable Energy sector?

Kind regards,

Concerned Citizen

Lincoln Coull

Then sequentially, I had noticed that for many months US President Joe Biden didn't have the numbers he needed to get his Climate Change bill over the line and passed into law. The US Congress was kept at a deadlock. This was mainly due to Senator Joe Manchin's' reluctance to support it. I felt I couldn't ignore it any longer; thought I had to try to help change this situation.

So I decided to commit another one of my Saturdays to sending a similar but more explicit email out to the world and then to every US Senator and then to the US President himself on 23/07/2022. Although I took a couple of 10 or 15 minute breaks, I sat in front of my computer from 11am and didn't finish until just before midnight. There were more Senators than I expected, but I got it done.

23/07/2021 - This email sent to most Canadian, Japanese, UK, China, USA, French, Russian, Holland, German, Indian and Australian media. And to every US Senator and the President.

Act now or see the adaptation method fail and the resulting turmoil

Hi,

It was very encouraging to view the recent Australian 60 minutes segment (01/05/2022) showing Australian mining billionaire Andrew Forrest and Australian Tech billionaire Mike Cannon-Brookes both claiming we have the technology now to service the entire worlds' energy by renewable sources and it is economically viable and profitable.

Andrew Forrest quoted "we need to move the world on from fossil fuels" and intends to show the world that it can be done while incorporating the lowering of operating costs, increasing production and increasing profits.

His company ran a mass scientific analysis that showed there is enough green energy to supply the whole world without using any fossil fuel ever again.

Andrew boldly and correctly stated we must stop this world from cooking".

Additionally, Australian Tech Billionaire Mike Cannon-Brookes, says "we can be the largest exporter of energy; the renewable energy Superpower". He and Andrew are currently developing the "Sun Cable Project" to sell Singapore our renewable solar energy and deliver 20% of their energy needs.

Right now, we are witnessing unprecedented heat waves in northern hemisphere that have led to many countries in Europe; Italy, Germany, France, Greece, Portugal and Spain all suffering from wildfires. While the US and China are experiencing extreme heat waves.

This reminds me of the recent unheard of fires in constantly cold, wet states like Tasmania and Alaska, caused by unprecedented lighting storms. The unbelievable and unprecedented national Australian bushfires of 2020 and the more recent devastating record breaking flooding along Australia's east coast. As I write this, parts of the state of New South Wales are currently experiencing their fourth consecutive flood inside the last six months; each one higher that the last. Unheard of.

More importantly, it's evident now that the yearly examples of extreme climate change events we have been witnessing, e.g. the rising seas, record storms and floods, giant wildfires all happening around the world, **have stepped up a notch.**

All of these events and the near collapse of the Gulf Stream, is further evidence that we are indeed experiencing real direct adverse effects of Climate Change **right now.**

As Dr. Daniel Cohen stated on twitter recently – "For the next 30 years, we'll be locked in an epic struggle over $100+ trillion in investment in reconstructing the built environment amidst climate chaos. "

As journalist David Wallace-Wells, who has spent years researching this and produced a book on the subject titled " The Uninhabitable Earth Life After Warming" correctly states – "If we are truly to solve the Climate Crisis, it would mean closing all fossil fuel industry in relatively short order." He further states "we have to wipe fossil fuel infrastructure off the planet". He also correctly informs us that we, the tax payer, is globally subsidizing these fossil fuel company's to the point of $5.3 trillion dollars per year, at the expense of deaths in the millions globally from their pollution.

It's worth noting that countries like Sweden, Costa Rica, Nicaragua, Scotland, Germany, Uruguay, Denmark, Morocco and Kenya have all achieved impressive Renewable Energy targets. However, while most all countries have pledged Net Zero emission by around 2050, it's countries like France, Greenland, Denmark, Belize, Spain, Ireland that have legislated the ban on extraction and production of fossil fuels or new exploration of them.

The latest IPCC Report published on 9th of August 2021 further confirms that "Human-induced climate change is already affecting many weather and climate extremes in every region across the globe. Evidence of observed changes in extremes such as heatwaves, heavy precipitation, droughts, and tropical cyclones, and, in particular, their attribution to human influence, has strengthened since the Fifth Assessment Report".

UN Secretary-General Antonio Guterres said that the IPPC report was "code red for humanity". "This report must sound a death knell for coal and fossil fuels, before they destroy our planet," "Countries should also end all new fossil fuel exploration and production, and shift fossil fuel subsidies into renewable energy." Guterres said.

I have been putting this message out to the world for the last 14 years now; that Climate Change is here and we are starting to pay significantly now, in the way of billions of dollars and millions of lives lost....

I put it to you, will you act responsibly now, will you demand your elected Minister/Senator to:

- Implement legislation to cease the use of all fossil fuels?
- Implement legislation to cease the use of all tax payers' fossil fuel subsidies to the fossil fuel industry and re direct them to the Renewable Energy sector?

Kind regards,

Concerned Citizen

Lincoln Coull

A few days later I learn that on the following Wednesday night (the 27th of July), Senator Joe Manchin changes his mind, he decides to end the deadlock and support the Presidents' bill. On the 16th of August 2022, to my humble gratification, after having been successfully voted in by both houses of the US Congress, President Joe Biden signed into law the ground breaking "Inflation Reduction Act", that will pump $370 billion into tackling Climate Change and various Energy programs. Further to that, the Governor of California Gavin Newsom has just announced an executive order that from 2035, demands all new sales of vehicles in that state to be emission free. There's a lot of expectation that about 17 other states will follow and then the rest of America will also fall into line and legislate accordingly.

We are starting to see the ongoing ripple effect.

Ironically, the newly elected and current Australian Labor Government also legislated its own Climate Change laws, including its emissions targets, on 8th of September 2022.

The law states that Australia will reduce its carbon emissions by 43% on 2005 levels by 2030 and net zero emissions by 2050. Poignantly, this legislation also includes a Safeguard Mechanism that "regulates the emissions of Australia's 215 biggest polluting facilities, including a who's who of fossil fuel companies and big miners"[1].

Importantly, the Safeguard Mechanism ensures:

- "Putting a hard cap on total emissions under the Safeguard Mechanism to ensure new and expanded fossil fuel projects can't blow this carbon budget out and drive up national emissions
- Subjecting any new project that would add significantly to emissions under the Safeguard Mechanism to a rigorous assessment and acting on the findings of this
- Sending a clear signal that big polluters should genuinely cut their emissions, not just rely on offsets to account for their pollution"[2].

After over a decade (or decades) of inaction by both countries, simultaneously we are witnessing them finally enacting laws that will encourage and demand the system, both in the private and public sectors, to kick start the vital action required right now to combat the Climate Crisis.

This is truly very heartening and encouraging to perceive.

Mindful of the 2015 COP 21 Paris Agreement, now that America, the second highest CO2 emissions producer in the world, has passed this extraordinary bill, it will motivate many other countries to follow suit in seriously decreasing their emissions. Parts of Europe is following and I expect we will see the vast majority of the 195 countries of the world that haven't already implemented similar laws, to do so.

From my stance

From my stance;

- Workchoices, the IR laws which disadvantaged the common worker, were abolished in my country in 2007

- A new message of protection for the Children was established in Australia, resulting in the Royal Commission into Institutional Child Abuse in 2013

- The resurgence of a global Climate Change movement grew, which significantly contributed to the 2015 Paris Climate Agreement and further global action

I have listened to media personnel, actors, ministers and senators say my very words, and have lived to influence Hollywood, Governments, Prime Ministers, Presidents, a Royal family, the Pope and many more….

You see…. I didn't do this for me….

PRIME MINISTER
CANBERRA

Reference: CA10/98 - 99

Lincoln Coull
noblelinc__@iprimus.com.au

Dear Lincoln Coull

Thank you for taking the time to write. Your good wishes and your kind words of congratulations are greatly appreciated.

It is a great honour to lead the Australian Government in these challenging and exciting times and I look forward to working with the Australian people to continue to build on the solid foundations of our fair, strong and progressive nation.

Yours sincerely

Julia Gillard

FROM MY STANCE

PRIME MINISTER

Reference: MC20-026934

9 June 2020

Mr Lincoln Coull
noblelinc@yahoo.com

Dear Mr Coull

Thank you for your letter regarding climate change.

Australia is taking strong climate action as part of a coordinated global effort. We are committed to the Paris Agreement and have a proud history of meeting and beating our international climate change commitments – we are on track to beat our 2020 target by 411 million tonnes. We will do this by investing in technology, and without increasing taxes, putting upward pressure on electricity prices or compromising the jobs of rural and regional Australians.

Our Paris Target, reducing emissions by 26 to 28 per cent on 2005 levels by 2030, is a responsible one that makes a significant contribution to global climate action. It represents a halving of emissions per person in Australia, or a two-thirds reduction in emissions per unit of GDP.

The Australian Government has a suite of practical policies in place to meet our emissions reduction targets while maintaining a strong and prosperous economy.

At the centre of our policies is the $3.5 billion Climate Solutions Package, including a new $2 billion Climate Solutions Fund. This builds on our previous $2.55 billion Emissions Reductions Fund and will help businesses, communities and landholders to reduce emissions. So far we have secured more than 190 million tonnes of emissions reductions, of which over 80 per cent will be delivered by the agricultural and land sectors.

Australia is also on track for around one third of our electricity needs to be met by renewables in the early 2020s. Right now, an unprecedented wave of clean energy investment is underway in Australia and new records have been set.

The Clean Energy Finance Corporation, the world's most successful green bank, has mobilised over $20 billion in new investments in our economy. I am proud to say Australia has one of the highest rates of per capita investment in renewable energy technologies in the world. We also have the world's highest uptake of rooftop solar – one in five homes have solar on their roofs.

To support the transition to renewable energy, the Government is investing in the energy storage and infrastructure of the future. This includes a $1.38 billion equity investment in the Snowy 2.0 project, which will be the biggest battery in the Southern Hemisphere.

Australia is also making progress towards increasing our overall energy efficiency by 40 per cent by 2030. The Climate Solutions Package, announced in February 2019, provides measures to improve energy efficiency, such as over $85 million to improve energy efficiency in homes and buildings and lower energy bills.

Our resources are supporting the transition to lower emissions around the world. We are among the world's largest exporters of Liquefied Natural Gas and hold among the largest reserves of lithium and cobalt for batteries. Natural gas is making it possible for nations to transition to a reliable and low emissions electricity supply.

In 2019 we released the National Hydrogen Strategy, through which Australian governments and industry are working together to build our hydrogen industry. The Australian Government has already committed over $500 million towards a hydrogen future. Looking ahead, we are developing a National Electric Vehicle Strategy to ensure a planned and managed transition to new vehicle technology. By the end of 2020 we will also develop a long-term strategy to reduce emissions, like other parties to the Paris Agreement.

We are also considering our next steps in practical action to reduce emissions. Given the economic challenges before Australia and the world, we must be particularly mindful of the impact these decisions have on jobs and livelihoods. That is why in May 2020, the Government released the Technology Investment Roadmap for consultation, which sets out our strategic view to guide future investment in low emissions technologies.

The Roadmap will help Australia develop new technologies to support jobs growth, back new industries to help our regional communities and economies to prosper, and maintain our strong track record of reducing emissions. Through our technology investments, Australia's action can contribute to our economy, rather than hurting it. Thank you for taking the time to write to me on this important issue. Australia is making a strong contribution to the global effort of tackling climate change while ensuring we have a robust and resilience economy

Yours sincerely

SCOTT MORRISON

noblelinc__@iprimus.com.au

From: "The White House" <noreply-correspondence@whitehouse.gov>
Date: Saturday, 6 December 2014 8:42 AM
To: <noblelinc__@primus.com.au>
Subject: Response to Your Message

THE WHITE HOUSE
WASHINGTON

Dear Lincoln:

Thank you for writing. I was touched by your kind words, and I appreciate your support for our shared values.

I believe we all have the power to make the world we seek. Our diversity and differences, when joined together by a common set of ideals, make us stronger. If we hold firm to our principles and back our beliefs with courage and resolve, then hope will overcome fear, and freedom will continue to triumph over tyranny—because that is what forever stirs in the human heart.

Again, thank you for your thoughtful message. I wish you all the best.

Sincerely,

Barack Obama

Facebook   Twitter   YouTube   Flickr   iTunes

> **We passed the Inflation Reduction Act thanks to you, Lincoln** →
>
> President Joseph R. Biden Jr. <info@contact.joebiden.com>
> To: noblelinc@yahoo.com
>
>
>
> Yesterday, I signed the Inflation Reduction Act into law -- one of the most significant laws in recent history. The American people won and special interests lost.

I live on knowing I have contributed to the betterment of humankind and this world.

I haven't been able to stop caring and still I keep an eye on the national and international Climate Change situation.

My altruistic endeavour to protect Mother Earth and the People remains ongoing….

A final thought

I write about these things as I expect you identify with them; you understand and share my deep concern.

In the media, we increasingly see a gradual rise in positive acts of protest by certain groups around the world; by many people who think the same way. They innately feel compelled to act towards combating the Climate Catastrophe. I suspect hundreds of thousands, millions of selfless people around the globe have been seriously protesting for action against Climate Change for well over a decade now.

However, mindful that this July 2023, was officially named the hottest month ever recorded, "the world as a whole has warmed by around 1.3C since the pre industrial period (1850-1900)"[1], 82% of the world's total energy is still supplied by fossil fuels [2] and global CO2 levels are currently at about 418 ppm [3], when they need to be under 350 ppm, it's clear, still nowhere near enough is being done.

Realistically, I sense that although there are positive signs that we are heading in the right direction in tackling Climate Change, overall, the human race is still failing at our most sacred role, that of being good guardians - caretakers - of our Earth; our Mother Earth who sustains us and our way of living.

Importantly, please let me reiterate the following in the lead up to my suggestion:

The latest IPCC Report published on 9th of August 2021, confirms "Human-induced climate change is already affecting many weather and climate extremes in every region across the globe. Evidence of observed changes in extremes such as heatwaves, heavy precipitation, droughts, and tropical cyclones, and in particular, their attribution to human influence, has strengthened since the Fifth Assessment Report".

As Dr. Daniel Aldana Cohen stated – Assistant Professor of Sociology at the University of California, Berkeley and Director of the Socio-Spatial Climate Collaborative – on X (formerly Twitter) 06/07/2022, "For the next 30 years, we'll be locked in an epic struggle over $100+ trillion in investment in reconstructing the built environment amidst climate chaos."

Furthermore, as journalist David Wallace-Wells, who has spent years researching this and produced a book on the subject titled "The Uninhabitable Earth Life After Warming" correctly states – "If we are truly to solve the Climate Crisis, it would mean closing all fossil fuel industry in relatively short order." He further states "we have to wipe fossil fuel infrastructure off the planet". He also correctly informs us that we, the taxpayer, are globally subsidizing these fossil fuel companies to the point of $5.3 trillion dollars per year, at the expense of deaths in the millions globally from their pollution.

UN Secretary-General Antonio Guterres recently stated that the IPCC report was "code red for humanity". "This report must sound a death knell for coal and fossil fuels, before they destroy our planet," "Countries should also end all new fossil fuel exploration and production, and shift fossil fuel subsidies into renewable energy."

A FINAL THOUGHT

The Scientists have been unanimous for years now, time is literally running out, the window is closing…

I've been putting this message out to the world for the last 16 years; that Climate Change is here and we are starting to pay significantly now, in the way of billions of dollars and millions of lives lost…

Considering the last COP 28 meeting in Dubai in November 2023 resolved to deep emissions cuts and scaled-up finance, signalling the beginning of the end for fossil fuels, it fell short of agreeing to the actual phasing out of fossil fuels completely.

Therefore, it is necessary for us to demand our governments go all the way now and seal the deal.

I put it to you, what's stopping you from contributing toward improving and helping this our Mother Earth?

I'll leave you with this final thought. Mindful that some countries have actually legislated the end of fossil fuel use and with the above mentioned, what's stopping you from asking each of your countries' elected representatives what I have recently asked mine?

- **Please implement legislation to cease the use of all fossil fuels**
- **Please implement legislation to cease the use of all taxpayers' fossil fuel subsidies to the fossil fuel industry and re direct them to the Renewable Energy sector**

Concerned Citizen,
Lincoln Coull

PS: Innately, we all feel we want to live in a clean environment, don't we?

I sincerely thank you for your time and reading about my life and views.

Notes

Save our Culture

1. The Living Planet: New Worlds – 12th April 1984 – written and presented by David Attenborough

Gore

1. An Inconvenient Truth – 24th May 2006 – written by Al Gore, directed by Davis Guggenheim, produced by Laurie David, Lawrence Bender and Scott Z Burns

Save Mother Earth

1. United Nations - FCCC/CP/2015/L.9 – Framework Convention on Climate Change 12/12/2015 – Conference of the Parties – Twenty-first session – Paris, 30 November to 11 December 2015 - ADOPTION OF THE PARIS AGREEMENT

How we got here

1. Why Jimmy Carter's Loss in 1980 Was a Loss for the Planet – Climate Change Was on the Ballot With Jimmy Carter in 1980 -Though No One Knew It at the Time - Time Magazine – by Jonathon Alter – 29/09/2020
2. "Losing Earth: The Decade We Almost Stopped Climate Change" – by Nathaniel Rich and George Steimetz – The New York Times Magazine – 01/08/2018.
3. The 1977 White House climate memo that should have changed the world - The Guardian, 14/06/2022, Emma Pattee
4. The 1977 White House climate memo that should have changed the world - The Guardian, 14/06/2022, Emma Pattee

5. "Losing Earth: The Decade We Almost Stopped Climate Change" – by Nathaniel Rich and George Steimetz – The New York Times Magazine – 01/08/2018.
6. Exxon Knew about Climate Change almost 40 years ago – by Shannon Hall 26/10/2015 – Scientific American
7. Exxon Knew about Climate Change almost 40 years ago – by Shannon Hall 26/10/2015 – Scientific American

Positive Reflections and Solutions - What we are doing

1. Why Australia's richest men are tackling climate change - Australian 60 minutes segment (01/05/2022)
2. Top 10: Nations that are leading the renewable energy charge – by Blaise Hope 16/03/2022 – https://sustainabilitymag.com
3. South Australian Government Finance Authority – https://www.safa.sa.gov.au
4. The end of fossil fuels: Which countries have banned exploration and extraction? – by Rosie Frost – 14/01/2022 - euronews.green – www.euronews.com
5. 2040 – Australian Documentary – Directed and starring Damon Gameau – Produced by Nick Batzias, Damon Gameau, Virginia Whitwell – released 11/02/2019
6. How to Save Our Forests and Rewild Our Planet – David Attenborough documentary – YouTube - WWF International – 02/05/2019 – https://youtu.be/lg9Tfc_hNsE
7. Costa Rica: the living Eden' designing a template for a cleaner, carbon-free world – 20/09/2019 – UN environment programme – http://www.unep.org
8. European Commission - Hydrogen buses: zero emissions, many benefits by Janusz Mizerny - https://ec.europa.eu/regional_policy/blog/detail.cfm?id=11
 D 5.3 – CHIC Final Report – 28/02/2017 –
 https://fuelcellbuses.eu/sites/default/files/documents/Final%20Report_CHIC_28022017_Final_Public.pdf
9. ITPro.Microsoft successfully tests emission-free hydrogen fuel cell system for data centres - by: Zach Marzouk - 29 Jul 2022
10. U.S. News - Clean Hydrogen Use Gaining Momentum in Countries By Elliott Davis Jr. - Dec. 27, 2021 - https://www.usnews.com/news/best- countries/articles/2021-12-27/hydrogen-use-gaining-momentum-in-countries-amid-c limate-change

NOTES

Forbes - Sustainability - Green Hydrogen, The Fuel Of The Future, Set For 50-Fold Expansion - By Mike Scott - Dec 14, 2020,10:00am EST - https://www.forbes.com/sites/mikescott/2020/12/14/green-hydrogen-the-fuel-of-the-future-set-for-50-fold-expansion/?sh=6484e8d06df3

We may have turned the corner?

1. Climate Council - online article - Safeguard Mechanism Decision Explainer – 27/03/2023 – reference to CEO Amanda McKenzie's twitter blog
2. Climate Council - online article - Safeguard Mechanism Decision Explainer – 27/03/2023 – reference to CEO Amanda McKenzie's twitter blog

A final thought

1. Carbon Brief – State of the Climate: How the world warmed in 2021 – Zeke Hausfather 17/01/22
2. Reuters – Renewables growth did not dent fossil fuel dominance in 2022, reports says – 26/06/23 – Shadia Nasralla
3. CO2.Earth – https://www.co2.earth > daily-co2 – 02/11/2023

Index

5th dimensional 30,
2030 48, 49, 56
2040 49,
2050 49, 56,
60 Minutes 47,
A1 29,
Aarau 51,
A New York Treat 25,
Adelaide 1, 12, 23,
Africa 33,
Agreement 19, 30, 56, 57,
AGL 49,
Agro Forestry 50,
Alaska 45,
Alexandria Ocasio-Cortez 19,
Altruistic 30, 62,
America 6, 32, 40, 42, 55, 56,
Amnesty 44,
Andrew Forrest 47, 53, 54,
Anthropocene 44,
Anthropologist 32, 33,
Australia 5, 6, 7, 45, 48, 49, 55, 56, 57,
Australian Culture 22,
Australian Government 5,
Arch Angel Michael 30,
Avatar 19,
Bangladesh 49,
Batteries 51,
Beijing 51,
Belize 49,
Bert Newton 7,
Billionaire 5, 47, 48,

Bill McKibben 19,
Bolzano 51,
Britain 43,
California 19, 55, 64,
Canada 8,
Canberra 10,
Capitalism 37, 43,
Carbon emission 8, 33, 40, 42, 56,
Charlotte Pass 11,
Child Sex Abuse 21, 22,
China 8, 42, 44,
CIA 40,
CICERO 8,
Clean Air Act 40,
Clean Water Act 40,
Climate Ad Project 8,
Climate Catastrophe 40, 43, 52, 63,
Climate Central 8,
Climate Change 8, 9, 11, 15, 19, 22, 37, 38, 40, 42, 43, 44, 45, 47, 52, 55, 62, 63, 64, 65,
CO_2 40, 43, 56, 63,
CO_2 emissions 40, 56,
Code red for humanity 55, 64,
Cologne 51,
Confucianism 26,
Concerned citizen 7, 53, 55,
COP 21 Paris 19, 51, 56,
COP 28 65,
Corporate 46,
Costa Rica 49, 50,
Covid 45, 47,
Crown 20, 21,

Damon Gameau 49,
David Wallace-Wells 64,
Deli Lama 25,
Democracy 4,
Denmark 49,
Doc Martin 11,
Domestic violence 45,
Dr. Brian Von Herzen 50,
Dr. Daniel Aldana Cohen 64,
Dubia 65,
Earth 8, 10, 11, 19, 27, 35, 40, 43, 51, 62, 63, 64, 65,
Ecosystem 50,
Electrolyser 48,
Email 6, 8, 12, 15, 19, 21, 29, 35, 52, 54,
Environ News 8,
Environment 15, 42, 46, 51, 64, 65,
Environmental 8, 40,
Environmentalism 3, 8,
Environmental legislation 40,
Europe 37, 44, 49, 56,
Extinction Rebellion 19,
Exxon Mobil 40, 42,
Feudalism 37,
Fifth Assessment Report 64,
Fiji 31,
Flag 11, 27, 28, 35,
Flinders University 3,
Fossil fuels 37, 42, 47, 49, 51, 63, 64, 65,
Fortescue Future Industries 48, 53, 54,
France 8, 44, 49,
Fulfillment 34,
Future Earth 8,
George Floyd 45,
Geostorm 19,
Germany 8, 44, 49,

GFC 43,
Gladston 48,
Global 2000 Report 40,
Global Warming 38, 42, 50,
God 4, 9, 17, 20, 21,
Gore 8, 9, 19,
Governor Gavin Newsom 55,
Graphs 16, 17, 38, 39,
Greece 4, 44,
Green energy 48,
Greenhouse gas 38, 42, 50,
Greenland 49,
Green Hydrogen 48, 49,
Green Peace 8,
Green Renewable 46, 49,
Greta Thunberg 19, 33,
Guardian 19, 63,
Gulf Stream 45,
Hamburg 51,
Heroes 23,
Himalayan 33,
Holland 8,
Hong Kong 44,
How we got here 37,
Humankind 19, 27, 62,
Human race 37, 63,
Human Rights 33, 34,
Humanitarians 25,
Hydrogen 48, 49, 51,
ICCCAD 8,
Idealism 32, 33,
Independents 6,
India 8, 23, 42,
Indian gentleman 23,
Industrial Revolution 37,
Inflation Reduction Act 55,
IPCC 8, 39, 64,
IR Laws 6, 7, 57,
Ireland 49,

Italy 44,
Japan 8,
Kenya 49,
Labor Party 6, 7,
Labor Government 52, 55,
Lake Jindabyne 10,
Lawyers 3,
Leonardo DiCaprio 19,
Letters 5, 29,
Liberal Government 6,
London 51,
Losing Earth 40,
Madam 29,
Madame Tussauds 25,
Mahatma Ghandi 25,
Marine Permaculture 50,
Martin Luther King 25,
Mateship 5,
Matriculation 3,
Medieval Europe 37,
Methane 50,
Michelle Obama 25,
Micro grid 49,
Mike Cannon-Brookes 47, 48, 53, 54,
Milan 51,
Minister 6, 19, 21, 52, 57,
Morals 34,
Morocco 49,
Mother Earth 10, 11, 19, 35, 62, 63,
Mother Teresa 25,
Mount Kosciusko 10,
Murray River 12,
Myanmar 44,
Naomi Klein 19,
Nathaniel Rich 49,
Natural Disasters 16, 17, 19,
Nelson Mandela 27,
Neo Liberalism 43,

Net Zero Emissions 49, 56,
New South Wales 45,
New York 25, 33,
NGO's 44,
Nicaragua 49,
No more Child Sex Abuse 21,
No more War 21,
Nonchalant Hero 23,
Oslo 51,
Paul Hawkens 50,
Peace 8, 26, 32, 33, 34,
Perth 51,
Phileas Cologne 51,
Politically aware 3,
Politician 3, 5, 8, 15,
Pope Francis 25, 35,
Portugal 44,
Positive Reflections 47,
Premier Peter Malinauskas 51,
President Clinton 33,
President Jimmy Carter 40,
President Joe Biden 53, 55,
President Lincoln 25,
President Obama 15, 19, 25,
President Reagan 43,
Prime Minister Julia Gillard 22, 58,
Prime Minister Scott Morrison 60,
Project Drawdown 50,
Protector 35,
Psychology 2,
Queensland Government 48,
Red Cross 44,
Religion 33,
Renewable 46, 47, 48, 49, 64, 65,
Republican 37, 43,
Respect 9, 29, 34,
Rock star 5,
Roman 37,
Royal Commission 21, 22, 57,

INDEX

Royal family 57,
Russia 8,
Safeguard Mechanism 56,
Save our Culture 5, 7,
Save Mother Earth 10, 11, 13, 35,
Scotland 49,
Seamans Hut 11,
Seaweed 50,
Senator 19, 52, 53, 54, 55, 57,
Senator Bernie Sanders 19,
Senator Joe Manchin 53, 55,
Sherpas 23,
Singapore 48,
Sir David Attenborough 8, 50,
Sir Edmund Hillary 23,
Solar 40, 48, 49,
Soleshare box 49,
South Australia 49, 51,
Southern America 32,
Spain 44, 49,
Spiritually 35,
Stockholm 50,
Summit 11,
Sun Cable Project 48,
Sunrise Movement 8,
Superpower 48,
Sweden 49, 50
Syria 44,
Tarot reading 30, 31,
Tasmania 45,
The Commission 21,
The Contemporary Anthropocene era 44,
The Day the Earth Stood Still 19,
The Flag 27,
The Greens 6,
The Guardian 19,

The US 5, 15, 19, 42, 44, 46, 53, 54, 55,
Tomorrowland 19,
Trade 32, 33,
Twitter 64,
Uluru 11,
UK 8,
UN 44, 55, 64,
Unicef 44,
University 3, 4, 5, 8, 26, 33, 64,
University of California, Berkley 64,
UN Secretary-General Antonio Guterres 55, 64,
US 5, 15, 19, 42, 44, 46, 53, 54, 55,
US Minister Alexandria Ocasio-Cortez 19,
USA 8,
US Senator Bernie Sanders 19,
Uruguay 49,
Vehicle Builders Union 3,
Victimization 44,
Virtue 34,
Whistler 51,
White House 40,
White House Council on Environmental Quality 40,
William Nierenberg 42,
Wind 49, 51,
Wodaabe tribe 33,
World Trade Organization 33,
Workchoices 57,
Workers rights 3,
Worship 33,
Yemen 44,
YouTube 12, 30,

73

www.ingramcontent.com/pod-product-compliance
Lightning Source LLC
Chambersburg PA
CBHW041319110526
44591CB00021B/2845